LAB LOGISTICS AND SAFETY

LET'S EXPERIMENT!

THE SCIENTIFIC METHOD IN THE LAB

ALISON AND STEPHEN ELDRIDGE

PowerKiDS press
New York

Published in 2021 by The Rosen Publishing Group, Inc.
29 East 21st Street, New York, NY 10010

First Edition

Editor: Jane Katirgis
Book Design: Reann Nye

Photo Credits: Cover Vietnam Stock Images/Shutterstock.com; series art Triff/Shutterstock.com; p. 5 Monkey Business Images/Shutterstock.com; p. 6 Kallayanee Naloka/Shutterstock.com; p. 7 https://commons.wikimedia.org/wiki/File:Synthetic_Production_of_Penicillin_TR1468.jpg; pp. 9, 13 Hero Images/Getty Images; p. 10 lcrms/Shutterstock.com; p. 11 Syda Productions/Shutterstock.com; p. 15 wavebreakmedia/Shutterstock.com; p. 17 Klaus Vedfelt/DigitalVision/Getty Images; p. 19 Compassionate Eye Foundation/Steven Errico/DigitalVision/Getty Images; p. 21 New Africa/Shutterstock.com; p. 22 Portra/DigitalVision/Getty Images.

Cataloging-in-Publication Data

Names: Eldridge, Alison. | Eldridge, Stephen.
Title: Let's experiment! the scientific method in the lab / Alison and Stephen Eldridge.
Description: New York : PowerKids Press, 2021. | Series: Lab logistics and safety | Includes glossary and index.
Identifiers: ISBN 9781725310421 (pbk.) | ISBN 9781725310445 (library bound) | ISBN 9781725310438 (6 pack)
Subjects: LCSH: Science--Methodology--Juvenile literature. | Science--Experiments--Juvenile literature.
Classification: LCC Q175.2 E43 2021 | DDC 507.2'1--dc23

Manufactured in the United States of America

CPSIA Compliance Information: Batch #CSPK20. For Further Information contact Rosen Publishing, New York, New York at 1-800-237-9932.

CONTENTS

WHY DO WE NEED A METHOD?

The scientific method is a set of steps scientists use to do **research**. For hundreds of years, it's helped scientists ask questions and make discoveries. But why do we need to follow a method when we do science?

The scientific method helps scientists gather facts to prove whether an idea is true. Using the scientific method, scientists come up with ideas and then test those ideas by observing facts. The scientific method even helps scientists check each other's work! They can follow the same steps to see whether they get the same results. This leads to better scientific **evidence**.

IN THE LAB!

In 1620, Francis Bacon of England created a method for scientific research. He didn't invent the modern scientific method, but he led others to create their own methods.

The scientific method is a series of steps scientists should follow to test their ideas.

ASK A QUESTION

Science begins with curiosity. When scientists set out to do research, they must have a question they hope to answer.

Usually, questions come from observations about the world. In 1928, scientist Alexander Fleming noticed that **mold** had **contaminated** one of his samples of bacteria. The bacteria couldn't grow near the mold. This observation led Fleming to wonder why. And that question led to the first experiments with antibiotics, medicines that have saved millions of lives.

Any question about the natural world can be a scientific question. Writing the question down helps scientists direct their research.

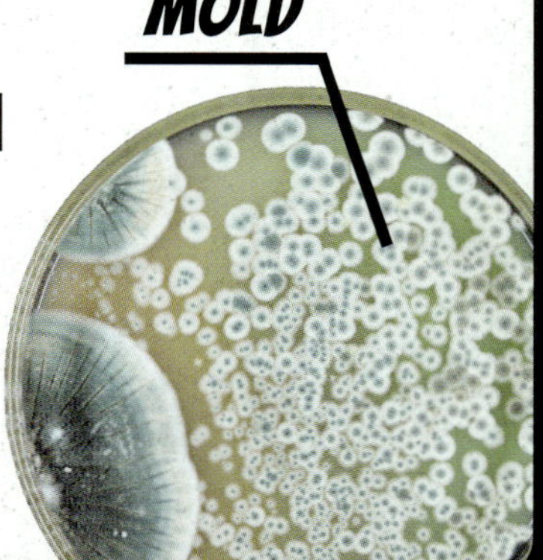

The type of mold Alexander Fleming discovered led to the creation of the medicine we know today as penicillin.

RESEARCH

Once you know the question you want to answer, the next step is to do research. Find out as much as you can about the subject.

Suppose you want to answer the question, "Does music help flowers grow?" You'll want to learn as much as you can about flowers and music.

Often, another scientist will have considered the same ideas you have. By reading their work, you can come up with new questions and ideas. If research suggests that plants grow better when exposed to music, then you can ask, "Do plants grow better with all sounds or only music?"

IN THE LAB!

Scientists often publish their work in magazines called scientific journals. These journals share important science with the world. They are very useful for scientists doing their own research.

Research your idea to see whether other scientists have made any discoveries about it.

CONSTRUCT A HYPOTHESIS

A **hypothesis** is a kind of educated guess that answers a scientific question. It's one possible answer, not necessarily the right one.

A good hypothesis allows you to make a **prediction** you can test. Suppose your hypothesis is that flowers grow better with noise. You can make a prediction that says, "If I grow flowers and play a noisy sound next to them, they'll grow faster than flowers that grow in the quiet."

A hypothesis that doesn't let you make a prediction is hard to test. If you can't test a hypothesis, you can't gather evidence to support or **reject** it.

A hypothesis is an educated guess. It's a possible answer you keep in mind while doing tests.

DO AN EXPERIMENT

Once you have a good hypothesis, it's time to test it with a scientific experiment. Experiments provide better evidence when they have fewer variables, or elements of the experiment that could change.

For example, suppose you want to test your prediction that flowers grow better with noise. You'll need more than one group of flowers, and you'll want to make sure the presence of noise is the only difference between the groups. If you use different kinds of flowers from one group to another or water one group more than the others, your evidence will be questionable. You can't know what variable caused the difference.

IN THE LAB!

The subject of an experiment is called an experiment group. It's compared to a **control group** to see whether the variable being tested had an effect.

Conduct your test and keep track of what happens.

ANALYZE DATA

During a test, scientists record observations about the progress of the experiment. These collected observations are called data. In an experiment that has to do with flowers, the data collected might be the height or health of the flowers over time.

Scientists then **analyze** their data. They try to understand what the information means. Sometimes they use mathematical tools to find hidden meaning in the data.

They may also try to make the data clearer using new ways to look at it, such as tables and pictures called graphs. These tools can help reveal evidence to support or reject the scientist's hypothesis.

Scientists use lots of methods to analyze data. A computer can be a handy tool to store and analyze data.

DRAW CONCLUSIONS

Once scientists have analyzed their data, they can use it to draw conclusions. Did the results of the experiment match the prediction? If so, that's evidence that the hypothesis is correct. If not, then the hypothesis may need to be changed or abandoned.

Sometimes experiments prove common beliefs wrong. Many years ago, people believed that some animals could just appear without being born from a parent animal. An Italian scientist named Francesco Redi did experiments to test the commonly held idea that rotting meat could turn into **maggots**. The results showed this idea was wrong. Today, we know all life-forms come from other life.

When you draw conclusions, you determine whether your hypothesis was correct.

COMMUNICATE RESULTS

After you've reached your conclusions, it's time to communicate your results. That means sharing your conclusions with the world.

Sharing information is one of the most important parts of science. Sometimes scientists communicate in scientific journals. Sometimes they talk to each other directly or give speeches at large meetings. Sometimes they talk to reporters about their findings.

Communicating results helps other scientists do their own research. Plus, when scientific knowledge is shared, it can be used! Many people use scientific knowledge in their jobs. The more we share knowledge, the more it can help people.

IN THE LAB!

When scientists publish their work, they include lists of other work that helped them. These are called citations. They give credit, or attention, to scientists who share important results.

Sharing conclusions helps other scientists do their own research and experiments.

DO IT AGAIN!

Good science doesn't end when results are published. When scientists share their work, other scientists will try to do it over again to check their results. In this way, we make sure the evidence for a hypothesis is strong.

The scientific method doesn't end. When one hypothesis is rejected by a test, someone will come up with a new hypothesis. When a test gives evidence for a hypothesis, someone will find another way to test it.

Every time a scientist shares data, it helps other scientists. Answers lead to more questions. Curious people will never run out of things to learn about.

IN THE LAB!

A hypothesis that has been tested many times can grow into a theory. In science, a theory is an explanation with very strong evidence to support it.

Answers lead to more questions, and more questions lead to more experiments!

ONE METHOD FOR EVERY QUESTION

The scientific method is a very **flexible** tool. It isn't a **specific** recipe for science experiments. It's a way of thinking about how we learn things.

Because of that flexibility, these steps can be applied to any question about the natural world. No matter what your question is, you can research it. You can make a hypothesis to answer it. If it's a good hypothesis, you can think of a way to test it. And you can analyze the results of your test to draw conclusions and share your findings.

Now that you know how to think like a scientist, start asking some questions!

GLOSSARY

analyze: To study something closely and carefully.

contaminate: To pollute.

control group: A group in an experiment in which the variable being tested can't have an effect on the results.

evidence: Something that shows that something else is true.

flexible: Easily changed or able to change.

hypothesis: A prediction about the answer to a question.

maggot: A wormlike insect that's the young form of a fly.

mold: A substance that grows on the surface of damp or rotting things.

prediction: A guess about what will happen in the future based on facts or knowledge.

reject: To refuse to accept, to prove otherwise.

research: Careful study done to find and report facts about something.

specific: Clearly stated.

INDEX

WEBSITES

Due to the changing nature of Internet links, PowerKids Press has developed an online list of websites related to the subject of this book. This site is updated regularly. Please use this link to access the list: www.powerkidslinks.com/lls/scimethod